Cut By: ______

Trains

based on the Thames Television Series

by Michael Robertson

Illustrated by Peter North

LUTTERWORTH PRESS LONDON

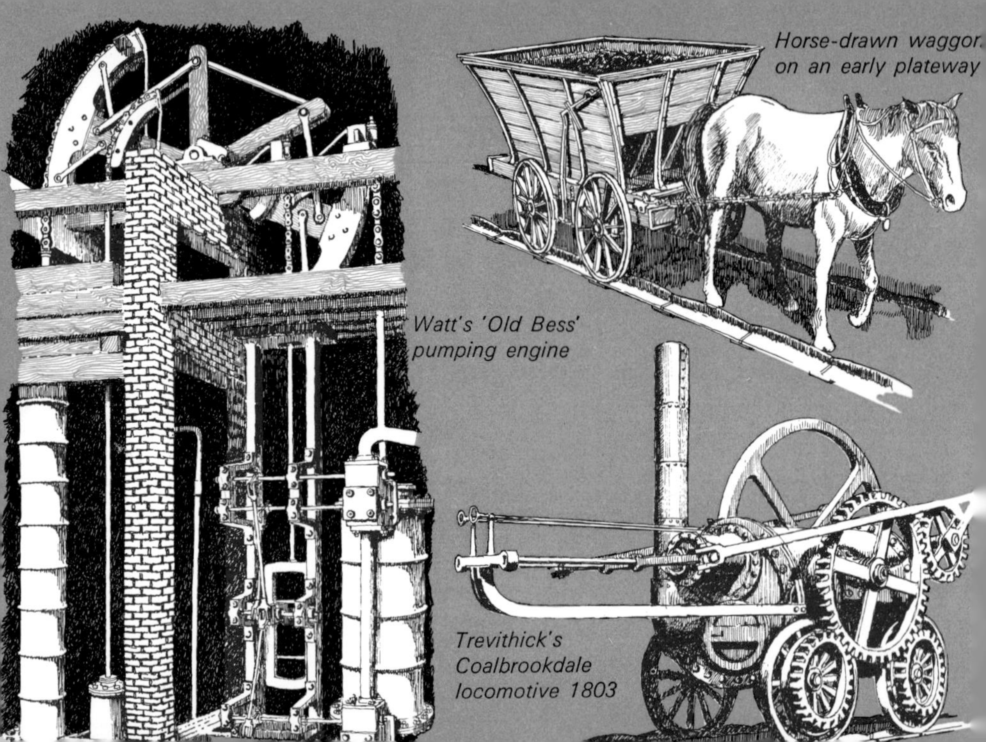

Horse-drawn waggon on an early plateway

Watt's 'Old Bess' pumping engine

Trevithick's Coalbrookdale locomotive 1803

how railways began

This is a book about railways and the huge, powerful, noisy, exciting engines which have pulled people and their belongings from place to place all over the country for so many years.

The first engines used on the railway were steam powered. James Watt built a successful steam engine in 1769. It hauled heavy loads up a hill towards it like a winch, but did not move. By this time tracks had been laid in mines along which horses could drag trucks of coal and rubble. At first these tracks were just flat metal plates until the idea came to use a track of raised rails to enable the trucks to roll along more easily. By the end of the 18th century steam locomotives had been built to travel on roads. The first to run on rails was probably the one built by Richard Trevithick in 1804 which was tested in Wales and found to be too heavy, breaking the rails it was running on.

John Blenkinsop and William Hedley improved the steam locomotive, but many people believed that the horse was a much more reliable and economic means of transport than the steam engine and they wanted them used on the railways. George Stephenson, the most famous name in railway history, changed all this. In 1825 his *Locomotion No. 1* ran the first public passenger steam railway service between Stockton and Darlington at speeds of 10 and 15 m.p.h. Horsemen who had come to gallop alongside could not keep up. Then, at a 'Contest of Speed' organised by the owners of the new Liverpool to

Trevithick's Pen-y-daren, 1804

Blenkinsop's Rack Locomotive. 1812

George Stephenson's 'Locomotion', 1825

Robert Stephenson's 'Rocket'
1829
ROCKET
Hedley's 'Puffing Billy'
1813
Kitson's long boiler
'Hector' 1845
HECTOR

An early train made up of first-class carriages and the 'Mail' hauled by 'Jupiter' on the

Manchester Line in 1829, Stephenson entered a locomotive called *Rocket* which travelled at over 30 m.p.h. The railways had well and truly begun.

Once it was realised that railways could do better than horse transport they spread rapidly, so much so that the years 1845–52 became known as the period of the Railway Mania! There was little organisation to decide where they should be built or who should build them. Rival lines were built in some areas. Locomotive improvement became very important. The steam engines got larger and more powerful. By 1850 speeds of 60 m.p.h. were being reached. Rothwell's 'Express Tank' locomotive was typical of the improved design. Engine drivers became almost rivals, coaxing the last ounce of speed from their trains. By

Liverpool and Manchester Railway, 1829.

the middle of the century there were at least thirty railway companies with train locomotives and trains, station staff and tickets. By 1872, the 393 miles from King's Cross to Edinburgh took only 9½ hours, including a 25-minute stop for lunch at York. This was more than four times as fast as it had taken by horse and carriage.

By 1888 competition between the London and North Western Railway and the Great Northern Railway grew so fierce on their routes to Scotland that they were racing each other to prove which had the better service. In the following years the battle continued, each company cutting its schedule by a few minutes in turn. Passengers and train enthusiasts alike loved these races.

Rothwell's 4 – 2 – 4 Express tank locomotive, 1853

Sterling's 8-foot Single, 1870

Drummond's 4 – 4 – 0, 1884

Dean's 4 – 4 – 0 'City of Truro', 1903

Ivatt's 4 – 4 – 2 'Atlantic', 1902

Churchward's 4 – 6 – 0 'Star' Class, 1907

the golden age

Among the locomotives which snorted and bustled our railways into the 20th century were the 'City Class', 'Star Class' and the 'Atlantics'. The Atlantics pulled the 'Queen of Scots' express, a non-stop train from King's Cross to Leeds, at maximum speeds of 90 m.p.h. To test his engine a driver once took a City Class locomotive from London to Plymouth at an average speed of 65 m.p.h. The Star Class was equally fast and important and one is still preserved at the railway museum in Swindon.

The early part of this century saw the railways at the height of their success. The fare from London to Glasgow was £2. The total length of railway routes was approaching the maximum in railway history. The *City of Truro* reached 104 m.p.h. in Somerset. During the First World War the railways suffered from lack of manpower, materials and money, and so ran far less efficiently. In 1923 the 123 small railways companies were combined by Parliament into four large companies controlling all the railways in Britain. These were the Great Western Railway, the London Midland & Scottish Railway, the London & North Eastern Railway and the Southern Railway. The disorganisation caused by the war did not stop the development of steam locomotives. In 1923 the G.W.R. produced the 'Castle Class' which from 1932 made the 'Cheltenham Flyer' the fastest train in the world, running from Swindon to Paddington at an average speed of 81.7 m.p.h. The 'King Class' of the same period on the

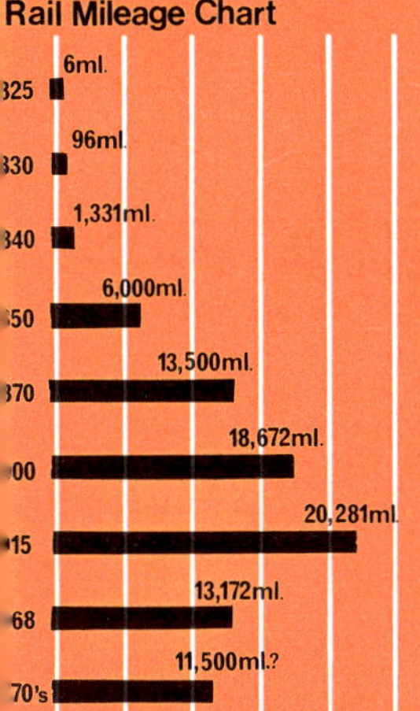

The steady expansion of track mileage which took place before World War 1 has been followed by a period of fairly rapid contraction. This has become much more marked in the last few years.

Western Region was another remarkable range of locomotives. In 1927 the 'Flying Scotsman' started making the longest non-stop service in the world from King's Cross to Edinburgh, pulled by Sir Arthur Gresley's 'Pacific Class' locomotives. The 1930s saw the fastest speeds that steam locomotion achieved. Gresley improved the Pacific Class to pull the London to Newcastle 'Silver Jubilee' train with the *Silver Link*. This train was streamlined to give it extra speed and balance, as was the 'Coronation Scot' pulled by Stanier 'Pacifics' between London and Glasgow. In 1938 a Gresley A.4 Pacific called *Mallard* reached 126 m.p.h. on a test – a world record for steam. Train services were at their peak for speed and this was truly the golden age of speed. The Second World War then intervened and again the railways suffered and the glories of steam were never recaptured.

After 1948 the Government took over the control of the railways. The accent was placed on modernisation, including the running down of steam power and its replacement by new forms of motive power.

The 'Coronation Scot,' 1937

The 'Silver Jubilee', 1935

The 'Brighton Belle' Pullman, 1939

The 'Cornish Riviera' 1935

The 'Golden Arrow', 1925

The 'Flying Scotsman', 1938

motive power

The steam engine has been by far the most exciting of our types of railway locomotives to see and hear at work. Now they are gone. They were dirty and noisy and much of their energy went up the chimney. Despite this, their evidence of giant power from clouds of belching steam and smoke made them very popular, and many people have mourned their passing. When the Southern Railway became very crowded with passengers and trains the change to electricity had to be made. Steam locomotives were marvellous on long runs at high speeds, but they took a long time to stop and start and on the crowded Southern Railway this counted against them.

There are two ways of using direct electric power on the railways. The Southern Railway decided to add a third rail to its tracks which carries the electricity and passes it on to the collector 'shoe' on the train. Most of the other regions which use

Clean and quiet, an overhead – electric Class A.L.6. locomotive on the London Midland region.
(*far left*) *Belching smoke, a Class B.1. engine hauls the up-line 'Fenman'.*

electricity have it passed to their locomotives by an overhead wire.

The rival to electric trains on today's railways is the diesel train. One kind is the powerful and complicated hydraulic diesels, of which the locomotive *Magpie* – in the 'Warship Class' on the Western Region – is an example. The most common diesel engine is more closely related to the electric engines and is called a diesel electric. These first ran on main lines of the Midland region about 20 years ago, their engines using diesel fuel to generate their own electricity, so turning them into travelling power houses. These are the engines which pull our trains today – clean, quiet and a long way from steam.

STEAM

STEAM
WATER
BOILER
SMOKE BOX
FIREBOX
PISTON

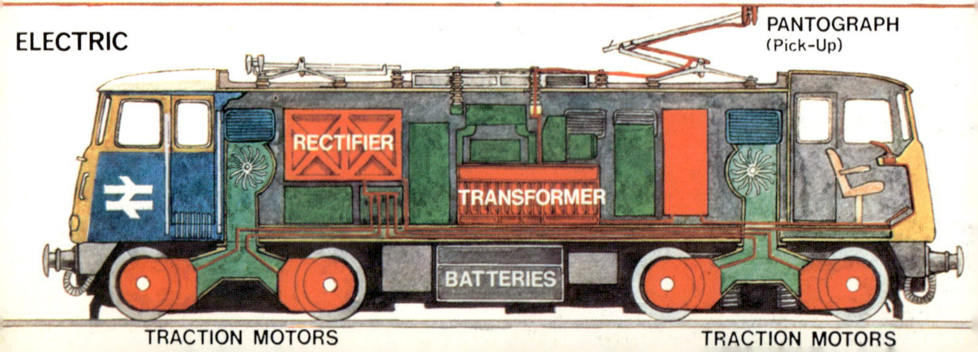

ELECTRIC

PANTOGRAPH
(Pick-Up)
RECTIFIER
TRANSFORMER
BATTERIES
TRACTION MOTORS
TRACTION MOTORS

DIESEL ELECTRIC

DIESEL HYDRAULIC (Magpie)

modern railways

Our railways have changed rapidly in the last ten years. Over 5,000 miles of track have been closed down as well as 1,500 stations. Lines had been built where there just were not the passengers to support them, and rather than drain the main lines of their profits to keep these branch lines open it was decided to use this money to improve the main-line services. Much of the Southern Region has been electrified on the third rail system since 1915, and the expense involved has been amply justified by the increasing use. Some 46,000 passengers pass through Waterloo Station at the peak of the morning rush hour. Over 2,000 trains pass through Clapham Junction every day. Apart from the huge number of electric multiple-unit trains carrying people backwards and forwards to work every day, Southern Region also runs diesel multiple units on the Charing Cross-Hastings line and the famous electric train, the Pullman 'Brighton Belle'.

The other electric-powered system, with overhead pick-up, is also now well developed. The A.L.6 electric expresses to Manchester and Liverpool on the Midland Region average over 70 m.p.h.

Diesel locomotives have also successfully taken over main-line services. Diesels arrived in growing numbers in the 1950s on British Railways, and on the Scottish and Western Regions in particular they have become well established. Both Regions use Sulzer-engined diesels, frequently coupling two together, especially on the Inverness-Perth Highland line. The locomotive peculiar to the Western Region is the diesel-hydraulic locomotive. Some of these have been adapted for working in pairs, when their 4,400 h.p. makes them the most powerful working unit on British Rail.

A mock-up of the interior of a new High-Density Passenger Coach at present being developed at the Derby works.

(far left). The Clacton main line on the Eastern Region of British Rail is served by these fast, overhead-electric multiple units.

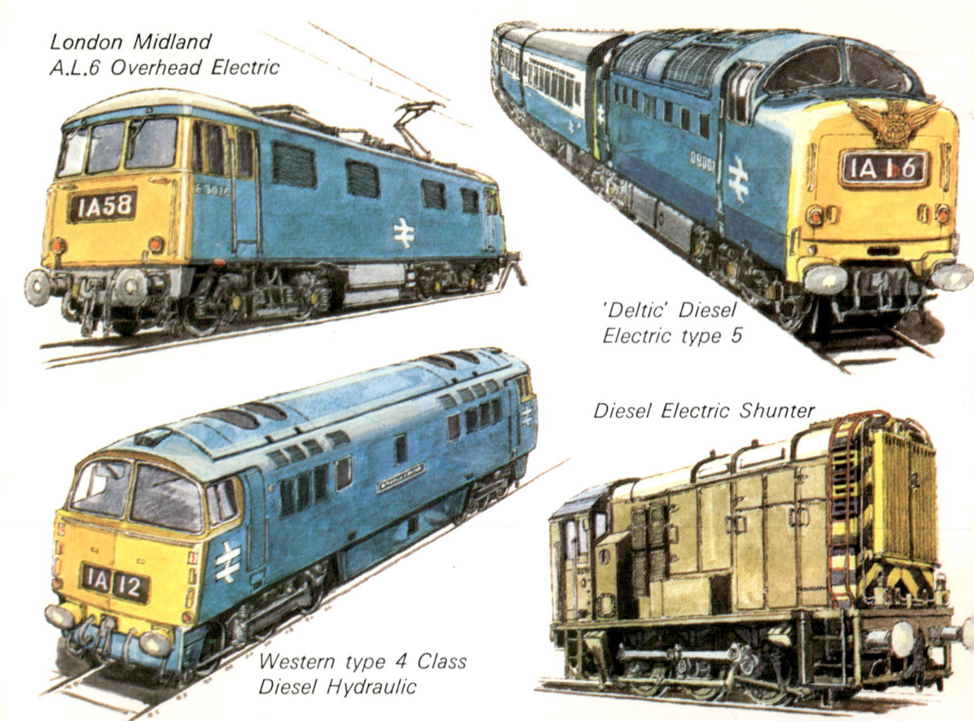

London Midland
A.L.6 Overhead Electric

'Deltic' Diesel
Electric type 5

Western type 4 Class
Diesel Hydraulic

Diesel Electric Shunter

Southern 3rd rail Electric Suburban Multiple unit

Southern D.C. 750 Electric Diesel

Western type 3 Class 'Heymek' Diesel

Diesel 'Pullman' Multiple unit

the travelling post office

The first Travelling Post Office van, complete with mail bag exchange apparatus, shows a close resemblance to the horse-drawn mail coach which it gradually replaced as the railways became established.

A present-day train picking up pouches from a wayside mail bag exchange apparatus.

The Travelling Post Office is an important part of our railway service. The T.P.O., as it is called, consists of four main Mail trains which run from Aberdeen and Penzance to London and in the other direction, and forty-five secondary T.P.O.'s made up of between one and six carriages which are attached to passenger or parcel trains and run all over the country. The letters they carry are sorted during the journey. Mail is collected and delivered at stations on the route without the train stopping. At night, at speeds of 40 m.p.h. and more, the T.P.O. men scoop up or drop off their mail bags. This requires a high degree of skill as they have to judge just by sounds and lights exactly where they are on what might well be a 400-mile journey, so that they can prepare for the next exchange of mail.

sleeping cars

The first sleeping cars appeared on London to Scotland expresses as far back as 1873, but up to 1928 these were only available to first-class passengers. Some lines used to provide sleeping carriages without any bed sheets, so you had to provide your own or go without. The modern second-class sleeper has two bunks placed one above the other. They are great fun to sleep in as you speed through the countryside and they are also very comfortable.

dining cars

The luxury of eating a meal while on your journey was first introduced to all classes of passenger in 1891. This caused extra work but it also increased the average speed of trains because they did not have to make the long stops for meals during the journey that had been a common feature of railway travel until this time. The kitchens on today's dining cars are very modern but extremely cramped. Imagine the problems of the chef preparing his menu in the narrow vibrating galley. Apart from the comfort and good food, some railway companies try to provide even more help for the passenger to forget his long ride, and on the trans-Australian trains a piano is included to allow people to sing and dance to pass the time.

Interior views of two of the various types of car used for the refreshment service of British Rail. (top) A Restaurant/Kitchen car (bottom) one of the new Griddle cars.

freight

Railways first began for the purpose of moving things, not people, and most of all minerals like coal and slate. Today the railways carry every type of material you can think of – food, oil, steel, cars, animals, everything. There are special trains for carrying these goods called freight trains. In Australia they have a freight train which carries nothing but sheep. It is a double-decker. In America freight trains give 'piggyback' rides to lorries and trucks to save them having to carry goods vast distances. In the same country triple-decker rolling stock is used to transport such things as cars, and trains can be up to two miles long.

The Goliath crane especially designed for the easy transfer of Container Freight from road to rail.

These roll-on, roll-off car-carrying wagons are seen here in use on the Motorail services. (below) A Freightliner made up of Container units.

In Britain the carriage of all types of goods is being much improved by the new freight liner system. Lorries, ships, aeroplanes, and trains are combining to transport goods quickly and easily. Special containers built for carrying one particular material or product are being made which can conveniently travel by all these types of transport. Cranes have been made which transfer the containers from ferry to train, or train to lorry. It will be easy to transport these containers to France or any other European country. Freight-liner trains can save up to two days by taking products distances of five hundred miles in one night and making their delivery possible by the next morning.

SPEED mph	10	20	30	40	50	60	70	80	90
'PUFFING BILLY' 10 mph									
'ROCKET' 30 mph									
'HECTOR' 50 mph									
'CITY OF TRURO' 102 mph									
'MALLARD' 126 mph									
'DELTIC' DIESEL 105 mph									
'OVERHEAD' ELECTRIC 110 mph									
mph	10	20	30	40	50	60	70	80	90

speed

Even today some of the speeds reached in Europe just before the Second World War are not being attained, although in general railway speeds are faster. At the moment there are maximum speeds of 90 m.p.h. on most of the British main lines and 85 m.p.h. on the Southern Region. This gives a safe breaking distance between signals at top speeds. As locomotive breaking power is improved, so speeds can go higher.

Nevertheless there are few locomotives in the world faster than the British 'Deltic'. In Germany and Russia there are a handful of trains which provide a similar service. But it is France which leads in European railway speed. One of their engines reached a test speed of 205.6 m.p.h. in 1955, although their daily working speeds are slightly lower than in Britain.

In Japan much inter-city track has been re-layed to give gentle curves for fast running and the Tokaido express averages 103 m.p.h.

maintenance

Trains are expensive vehicles and have to be looked after very carefully. The maintenance of their engines is particularly important. Passengers hate delays and breakdowns and the railway companies cannot afford to have their locomotives standing idle with mechanical faults. Each of the regions of British Railways has depots specially for repair work. These maintenance sheds are like a doctor's waiting room. The giant locomotives queue up for their inspection and when their turn comes they are taken to pieces and checked. Electric and diesel locomotives have separate sheds for repair work, as diesel engines are more

Re-wheeling a Southern Region electric locomotive M.11.

complicated and delicate than electric and need more attention.

The Laira depot at Plymouth is a good example of a modern maintenance yard for diesel locomotives. Their Western Region diesels travel between 8,000 and 9,000 miles every 24 days, after which they go to Laira for full maintenance. The mechanics work non-stop for 16 hours to check the locomotive before it is ready to return to service. This work has to be done as fast as possible so that the diesel can get back to work with the minimum delay. As the locomotive leaves Laira it gets a good wash and is then back to another 24 days hard work before the next check-up.

Next time you ride in a train just remember all the work involved.

A diesel locomotive engine being lowered into its body.

narrow gauge railways

Steam power can still be seen on Britain's only mountain 'rack' railway. This railway was built in 1897 and goes up Mount Snowdon in North Wales. The 'rack' consists of two rails with teeth which engage a wheel from the locomotive.

The majority of mountain railways are narrow gauge. The railway gauge is the distance between the two railway lines. In Britain there is a standard gauge used by British Railways of 4 ft. $8^1/_2$ in. However, there are a few railways which have narrower tracks and they are all called narrow gauge.

In Britain the first narrow gauge lines were built in North Wales to carry slate away from quarries. In 1836, in the same area, the Festiniog 2 ft. – gauge railway carried passengers, although it did not use steam power until 1863. Many other 'light' railways followed. Only the Vale of Rheidol is owned by British Railways; the others are all kept going privately.

Snowdon 'Rack' Mountain Railway

The following Narrow Gauge railway services are still in existence:

The Talyllyn.
The Festiniog.
The Welshpool and Llanfair.
The Vale of Rheidol (B.R.)
The Fairbourne Miniature Railway.
The Ravenglass and Eskdale.
The Romney, Hythe and Dymchurch.
The Isle of Man.
The Manx Electric.
The Lincolnshire Light Railway.
The Great Orme Railway.
The Volks Electric Railway (Brighton)
The Snowdon Mountain Railway.
The Smaefell Mountain Railway (Isle of Man).

The Festiniog's double ended 'Earl of Merioneth' (30 ft)

The Talyllyn Railway's original engine (16 ft 10 in.)

The Romney, Hythe and Dymchurch 'Hercules' (28 ft 2 in.)

international trains

As in Britain, steam has been abandoned by all the great railways in the world. The following are some of the most important and glamorous of international trains.

The new Tokaido line of Japan is a model for the future of railways. This line links many of the country's most important cities and is serviced by electric trains of high speed and frequency. The pride of the Italian railways is the 'Settebello', another inter-city train which successfully challenges the competition of domestic airlines. The driver is in the upper deck of the front carriage while the passengers can sit in the observation car in the nose of the train.

Russia uses many types of locomotive power for their trains: diesel, electric, the gas turbine

The super-expresses on the Tokaido line of the Japan National Railways are claimed to be the world's fastest trains. These overhead-electric multiple-unit trains cover the 320 miles between Tokyo and Osaka in 3 hours, with a maximum speed of 125 m.p.h.

engine, and even a few steam trains. The 'Red Arrow' is electric-powered and covers the October Line between Leningrad and Moscow each night. One of the fastest trains in the world is the French 'Mistral' Express. If you ever go on holiday to the Mediterranean you may well travel on the 'Mistral'. The Trans-Europ Express is a service made up of many famous electric and diesel trains which run between the major cities in Germany, France, Italy, Switzerland, Belgium and the Netherlands.

The South African railway is famous for its service and economy. The 'Orange Express' is one of its most important trains. Its overhead electric locomotives are very powerful and pull at least fourteen carriages on this service. One of America's most famous railways is The Pennsylvanian, where the trains are powered by diesels.

The famous 'Blue Train' of South African Railways.

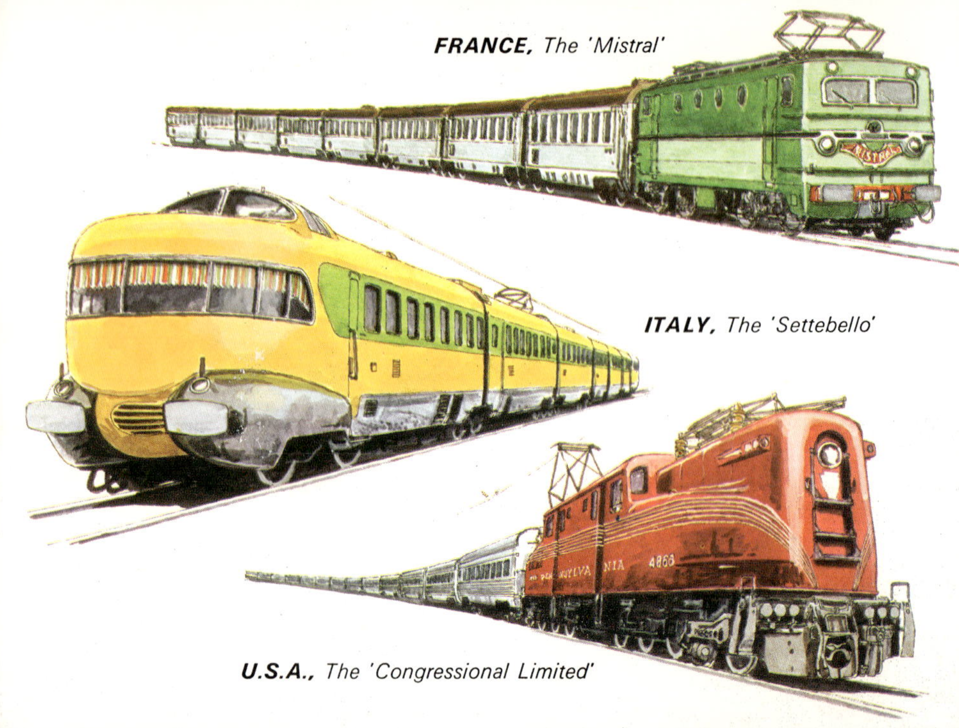

FRANCE, *The 'Mistral'*

ITALY, *The 'Settebello'*

U.S.A., *The 'Congressional Limited'*

SOUTH AFRICA,
The 'Orange Express'

CANADA,
The 'Canadian'

U.S.S.R., *The 'Red Arrow'*
Moscow to Leningrad

ferries

To make our travelling fast and convenient transport should be closely linked. To be able to change quickly from a bus to a train or aeroplane is as important on a journey as the speed of the vehicle itself. With this in mind our railways have always worked closely with sea transport. In fact British Railways have their own fleet of ferries with direct travel to Guernsey, Jersey, the Isle of Wight, Ireland, Holland, Belgium and France.

Today we call the trains which connect with the ferries, 'boat trains'. Two of the most famous are on the Southern Region: the 'Golden Arrow' and the 'Night Ferry' on the London to Paris service. This train is composed of Continental and Southern Region vehicles. The sleeping cars and brake van on the 'Night Ferry' travel across the Channel and on to Paris. On the return journey French rolling stock is hauled to Victoria.

future developments

The prototype of the new High-Density Passenger Coach now being tested at the Derby works.

Since the last war our railways have put on a bright modern face and are as exciting as they always were. Perhaps there is less individuality and character since the passing of the steam locomotive but who knows if one of the new developments may not produce a similar magnetism.

high density urban transport

One of the problems now facing the railway engineers is the design of a train to carry the increasing numbers of people who travel into the big cities to work each day. At Derby, the largest railway research centre in the world, a new passenger coach has been designed to answer this purpose.

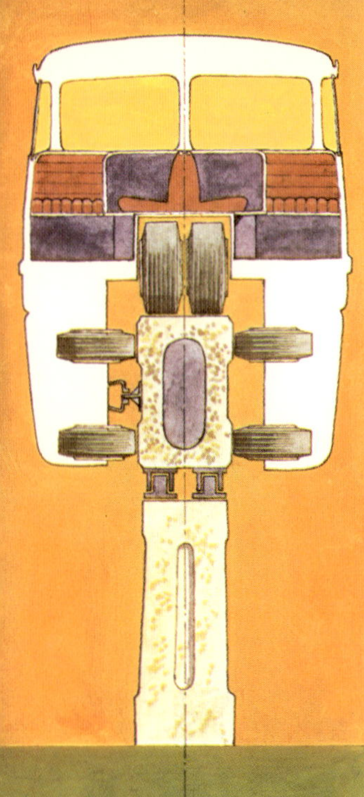

the monorail

Widely used transport of the future might be a Monorail train. Some of these are already in operation. The train may hang under the track, as used in a German system, or perch on top of the rail, as shown in the diagram on the left. They could be built above motorways, easing them of their increasingly heavy traffic without taking up additional land.

signalling

The experiments with different types of trains are intended to make the railways even faster than they are today. It is of course important that they should be safe, and that is why work is going ahead on a new signalling system which will keep the train driver in two-way communication with a central control all the time.

Riding on a single concrete beam, the Alweg Monorail System carries passengers in such far away places as Seattle, Washington; Disneyland, California; Tokyo, Japan; and, nearer home, at Turin in Italy (shown here).

the tracked hovercraft

Tracked Hovercraft Ltd., a subsidiary of the National Research Development Corporation, are developing a type of hovercraft which will travel over concrete track and could be operated as a form of railway. It is not intended to replace railways but to supplement some of their services. The Tracked Hovercraft could produce faster passenger services between city centres at speeds of between 200 and 400 m.p.h. One possible plan of a 100–150 seat vehicle for future passenger use is shown in this artist's impression. In this way they would certainly improve on our existing inter-city air transport. They could also be adapted for freight and local passenger traffic. They will travel along tracks about seven feet wide upon a cushion of air in the same way as the cross-Channel hovercraft travels over water.

advanced passenger train

One of the newest and most important projects of British Railways is for an Advanced Passenger Train. This will be powered by gas-turbine engines and promises to be a great advance in train history. A model of the train is shown in the picture. The aim of this train will be to make the ride as smooth as possible. The carriages will be of very light metal. The wheels will have special rubber mountings and the whole train will be designed to tilt over at an angle as it turns corners, rather as a motor cyclist does. In this way it is hoped that the train will be able to run at speeds up to 150 m.p.h., which means, for instance, that the journey from King's Cross to Newcastle, which today takes 3 hours 50 minutes, will be done in something like 2 hours 40 minutes.

The 'cab' of an overhead electric locomotive
1 *Vacuum air brake valve;* ***2*** *Straight air brake valve;* ***3*** *Automatic warning system;*
4 *Main air reservoir gauge;* ***5*** *Brake cylinder gauges;*
6 *Exhauster, Pantograph and train heat controls;* ***7*** *Train pipe/vacuum chamber gauges;* ***8*** *Speedomete*
9 *Traction motor ammeters;* ***10*** *Liveline, fault and train heat indicators;*
11 *Instrument light controls;* ***12*** *Speed controller;* ***13*** *Master control key;* ***14*** *Forwar reverse handle;*
15 *Rheostatic brake;*
16 *Horn;* ***17*** *Windscreen wiper control;*
18 *Anti-slip brake butt*